Metal Detecting Benefits for Landowners

David Villanueva & Jacq le Breton

CONTENTS

1 INTRODUCTION

Metal detecting first became popular as a hobby in the 1970's with affordable machines becoming available. Recreational sites such as beaches and woodland proved to be rich in finds. With unusual and sometimes valuable finds coming to light, attention turned to pottery fragments and metal objects uncovered by the ploughs. These were possible evidence of long lost settlements. War time relics unearthed determined battle sites or camps perhaps, while trading tokens might indicate ancient market sites. The keener detectorist began to acquire permissions on arable land and pasture, **free of crops**, with the landowner enjoying a share in both the making of history and financial reward.

As metal detectors became more advanced and the hobby grew, detecting clubs were formed together with national organizations: The National Council of Metal Detecting (NCMD) and The Federation of Independent Detectorists (FID). Both organizations recognized the need for members to have public liability insurance and currently provide £10,000,000 cover, protecting landowners in the unlikely event of damage or loss as a result of metal detecting activities.

In order to promote and define responsible metal detecting, a Code of Practice was drawn up, which has been slightly amended over time. The latest version, included in this booklet, has been agreed by all key archaeological bodies, together with metal detecting and landowners' organizations to provide a clear and unambiguous definition of what constitutes good practice.

General awareness of what lay under our feet has grown. Coins and artefacts lost, hidden or discarded can be found from the very latest to those dating back thousands of years. From time to time a hoard or item of Treasure turns up. *Imagine if this happened on your land!*

2 A BRIEF HISTORY OF METAL FINDS

Metal artefacts first appeared in the Bronze Age and have expanded in type and variety ever since. Typical dates and divisions of the periods are given below but bear in mind that the dates are generally approximations and there was a gradual change from one period to the next. The issue is also somewhat confused by the use of different date ranges and alternative names. The main divisions are the British Museum version but common alternative names and date ranges are given in brackets.

Bronze Age (Prehistoric): C. 2500 – 700 BC

Early: C. 2500 – 1500 BC; Middle: C. 1500 – 1150 BC; Late: C. 1150 – 700 BC

Metal finds of the Bronze Age will usually be copper-alloy (bronze) tools or weapons. There is also the possibility of gold jewelry in the form of torcs (twisted wire neck and arm rings) and high status gold vessels like the Ringlemere cup.

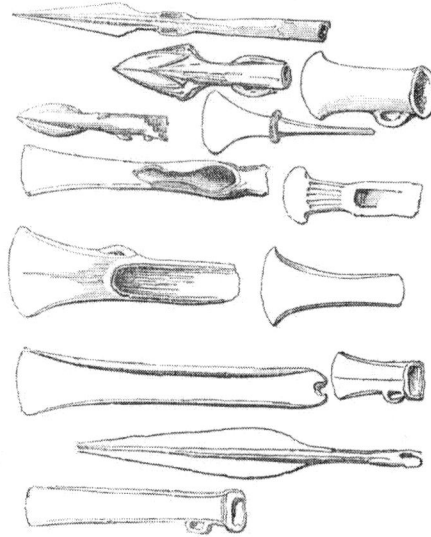

Bronze Age tools and weapons

Iron Age (Prehistoric, Celtic): C. 700 BC – 43 AD

Early: C. 700 – 350 BC; Middle: C. 350 – 100 BC; Late: C. 100 BC – 43 AD

Tools and weapons were generally made of iron but will not often be found owing to corrosion and metal detector discrimination settings. More often found will be fittings in bronze, such as knife handles and sword and dagger pommels.

Clothing fasteners in the form of brooches and toggles started to be used. Iron Age bow brooches were made of copper-alloy, which, in their simplest form, looked similar to a modern day safety pin, although there were annular brooches and the more ornamental dragonesque brooches. Copper-alloy pendants also appeared and high status adornment continued in the form of gold torcs and finger rings.

Copper-alloy mounts for attachment to leatherwork belts and buckets were used together with fittings associated with horses, such as strap junctions and terret rings. A unique cosmetic implement in the form of a woad grinder was introduced and hand-held mirrors are also occasionally found.

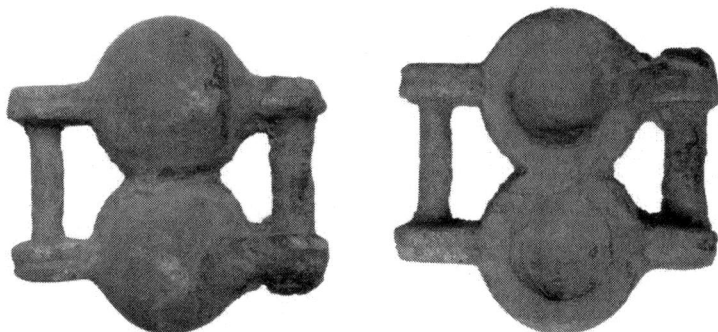

Iron Age strap junction

Although there is a school of thought that bronze axe heads were used as a type of early currency, iron bars and gold ring money was used in the Iron Age, developing into a coin based currency from the Middle Iron Age, with the introduction of a range of copper-alloy, silver and gold coins.

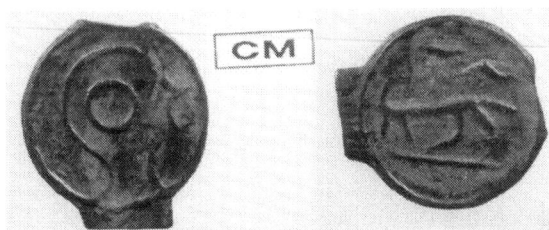

Iron Age *potin* coin, c. 50BC

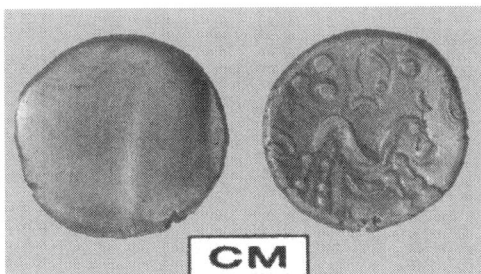

Gallic War gold *stater* coin, c. 50BC

Roman 43 – 410 AD

(Romano-British C. 100 BC – 100 AD)

The list of artefacts left in the soil by the Romans is quite enormous compared to previous ages and the range and variety expanded accordingly.

Tools and weapons generally continued to be made of iron, often with fittings in copper-alloy, such as knife handles and sword and dagger pommels. Iron horseshoes in the form of strap on hippo-sandals also appear.

Brooches developed into a great variety of more elaborate affairs, with flat plate and zoomorphic brooches coming into use. Most Roman brooches are made of copper-alloy, often with coloured enameling. Pendants continued, many of an erotic nature, which were made mainly in copper-alloy and occasionally silver or gold. Dress pins became fairly common.

Copper-alloy mounts for attachment to leatherwork, utensils and furniture were used, while figurines and small statues are sometimes found. Spoons made from copper-alloy or silver are occasionally found along with other domestic ware such as pans and plates (*pateras*). Finger rings are quite common in copper-alloy, silver and gold. Buckles, seal boxes and keys appeared as did steelyards and associated weights.

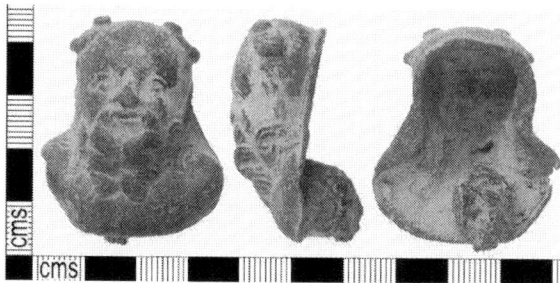

Roman furniture mount featuring Bacchus

Copper-alloy clapper bells, which are more or less conical, open at the widest end and fitted with a swinging arm or clapper, appear in

Roman times. Cosmetic or medical implements such as ear scoops and dental picks in copper-alloy or silver are also found.

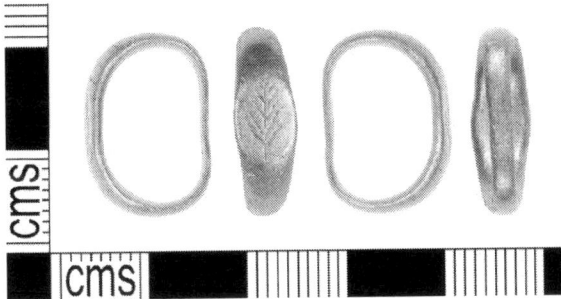

Roman gold finger ring

The Romans had a well developed coin based currency system, which changed frequently with the emperors, producing a huge range of copper-alloy, silver and gold coins.

Roman bronze coin of Fausta, 307-337

Silver *denarius* of Antoninus Pius, 138-161

Early Medieval (Saxon)

Early (Dark Ages): C. 410 – 720 AD; Middle (Dark Ages): C. 720 – 850 AD; Late (Viking): C. 850 – 1066 AD (Anglo-Norman: C. 1000 – 1100 AD)

Coins (mainly silver) and artefacts are quite scarce finds from this period and often the artefacts are of high quality and may be gilded or even solid silver or gold. Weapons and shield bosses are often associated with a male burial, whereas personal adornments may indicate a female burial.

Tools and weapons generally continued to be made of iron, often with fittings in copper-alloy, such as knife handles and sword and dagger pommels. Iron shield bosses are occasionally found. Iron Horseshoes take on the modern form but generally have a wavy edge. Horse harness fittings are found in the form of strap fittings and bridle cheek pieces as well as stirrup mounts. Belt and scabbard fittings include buckles, chapes, mounts and strap ends.

1: strap-end 2: Athelred penny 3: buckle

In the Early Medieval period bow brooches became even more elaborate; plate brooches developed into button and saucer brooches and annular (solid ring) or penannular (open ring) brooches appeared. Most brooches are copper-alloy but usually gilded and may feature enamel or stones such as garnets; high status brooches were made from silver or gold. There were a large variety of pendants in use in bronze, silver and gold. Finger rings, typically made from silver or gold twisted wire, are occasionally found. Hooked clothing fasteners, dress pins, keys and tweezers also turn up.

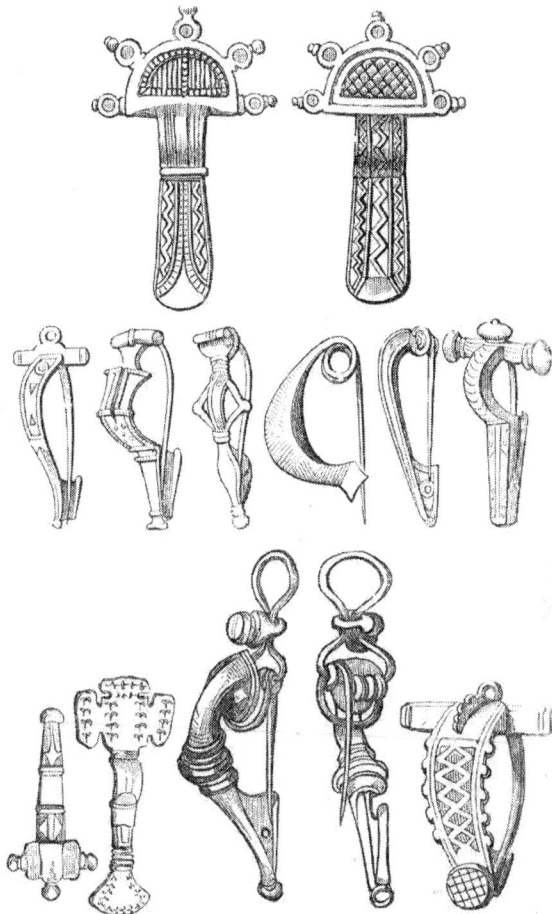

Early Medieval Brooches

Medieval C.1066 – 1500 AD

Tools and weapons generally continued to be made of iron, often with fittings in copper-alloy, such as knife handles and sword and dagger pommels. Horseshoes, made of iron, take on the modern form but usually are keyhole shaped. Horse harness fittings are generally confined to pendants, often heraldic, with enamel. Belt and scabbard fittings include buckles, chapes, mounts and strap ends.

Medieval brooches are usually annular, typically in copper-alloy but considerable numbers of silver and even gold versions are found. Pins in copper-alloy and silver are also found. Finger rings become more common.

The crotal or rumbler bell, which is usually spherical (sometimes pear shaped), enclosing a loose iron ball or pea to generate sound and having a suspension loop for attachment to (usually) animals, came in to regular use. These were mainly made in copper-alloy, tin or pewter and date to 13th – 15th centuries but may be as late as 19th century.

Lead seals and seal matrices in lead, copper-alloy and occasionally silver are fairly regular finds from this period. As well as personal seal matrices there are large circular lead seals, called Papal Bullae, originally attached to documents from the Vatican. Cloth, bag and sack seals in lead can also be found. Pilgrimages were popular during this period and result in finds of lead ampullae (for holy water) and pilgrims' badges.

Many of the artefacts that we are familiar with today came into regular use during this period:

* Buttons appeared in the 14th century and are cast in one-piece from copper-alloy or pewter.

* Change purses were made of leather or cloth but were suspended from a copper-alloy purse bar.

* Keys, many made of copper-alloy, are found quite regularly.

* Thimbles of copper-alloy, having a closed top, date from 1350.

* Weights for trade of copper-alloy and lead; coin weights of copper-alloy and lead loom weights and spindle whorls used for spinning and weaving.

**1: seal matrix 2: thimble 3: key 4: spur rowel 5: ring brooch
6: lead seal matrix**

232 and 233. Pence of William I. or II.   235. Angel of Henry VI.   237. Quarter-noble of Henry IV.
234. Penny of Edward I.   236. Shilling of Henry VII.   238. Gold penny of Henry III.

Medieval coins

The basic coin was the silver penny or sterling, which was divided into halfpennies and farthings by cutting the penny into fractions. From the 12th century round halfpennies were produced, followed by round farthings from the 13th century. The silver penny remained as the highest denomination until the silver groat of fourpence was coined from 1279 and halfgroat added in 1351. The silver coinage continued unaltered in range from farthing to groat until the last years of the medieval period when Henry VII added the testoon or shilling of twelve pence. Gold coinage in the form of the Noble (80 pennies), half-noble and quarter-noble was struck from 1344. In 1464 the Ryal or rose-noble of ten shillings, was introduced, with halves and quarters, and the noble was effectively renamed the Angel. Half-angels were struck from 1470. A Sovereign of 20 shillings was added in 1485.

There was no official base metal coinage in England during this period. Copper-alloy jettons were introduced for accounting and may have been used as small change, however.

Post Medieval C.1500 – 1800 AD

(Tudor: C.1500 – 1600 AD); (Stuart: C.1600 – 1700 AD); (Georgian (Hanoverian): 1715 – 1837)

As we head into the age of industry and invention there is a proliferation of both coinage and metal artefacts. A few highlights are:

* The table fork comes into use.

* Posy (the forerunner of wedding and engagement rings) and memorial or mourning rings in gold (mainly) and silver become popular.

* Firearms come into use with lead ammunition (musket and pistol balls). Toy pistols and canons are also found.

* Tobacco for smoking and snuff is imported from the colonies (America). Pipe tampers and other tools become popular.

* Thimbles with open tops, called sewing rings, appear.

* Fittings (mainly copper-alloy) become common on furniture.

* Post Medieval and Modern brooches tend to be much more ornate and come in a great variety of forms. Many will be silvered or gilded copper-alloy but there is an increased chance of finding solid gold and silver examples.

The method of coin production changes from hammered to milled (using a coin press). The increasing price of silver reduces the penny to a very small size and copper farthings and halfpennies replace the silver penny for small change. Shortage of small change generates periods of unofficial token production in lead (16th-17th centuries) and copper-alloy (17th -19th centuries).

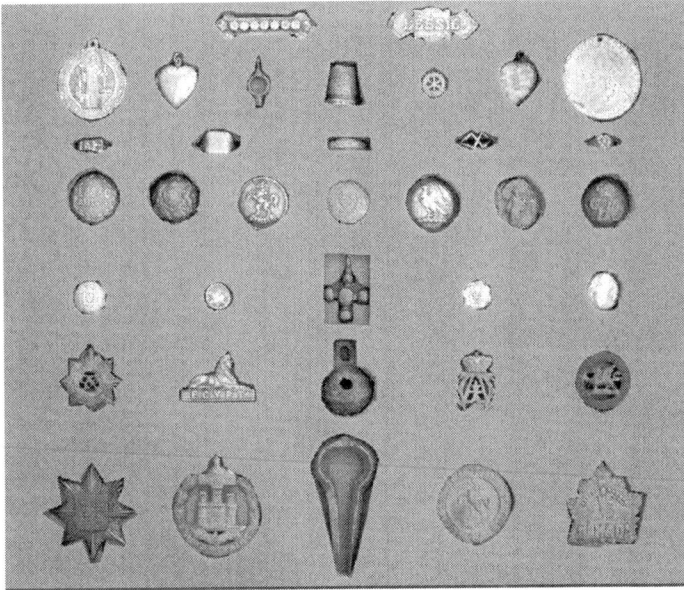

A range of Post Medieval and Modern finds

Modern 1800 AD – present

(Victorian 1837 – 1901)

Modern finds are much more common than finds from earlier periods principally because there were more people around who had metal objects to lose. Most modern metal objects are machine made and this shows up in the regular shape and design. Screw threads are largely a modern invention so if an artefact has a screw thread it will almost certainly be modern. Most horse furniture (brasses etc.) dates from Victorian to early 20th century. Railways came into being in the 1820s and the motor car in the 1890s so anything related to either can only be modern.

After a brief show of tokens in the early 19th century, there was a re-coinage resulting in coins being very much as they are today. However, during the 20th century gold and silver ceased to be used for coins in normal circulation to generally be replaced by nickel-brass and cupro-nickel. As a result of rising metal prices, lower value coins are increasingly being made from plated steel.

3 TREASURE AND HOARDS

It is only a matter of using a metal detector in the right place to turn up something special and that could be anywhere! While most of the non-junk objects we find are historically interesting pieces of scrap metal, every now and then something turns up that is much more than that. Single rare coins and artefacts can be worth hundreds or thousands of pounds in their own right but when hoards turn up they can lead to riches beyond the dreams of avarice, shared between finder and landowner. Tenants and occupiers could benefit too.

The Treasure Act applies to metallic objects at least 300 years old when found and defines **Treasure** as any single artefact of at least 10% by weight gold or silver. And all coins that contain at least 10% by weight of gold or silver coming from the same find consisting of at least two coins. And all coins containing less than 10% by weight of gold or silver coming from the same find consisting of at least ten coins. And any associated objects (e.g. a pot or other container), except unworked natural objects, found in the same place as treasure objects. And prehistoric (i.e. Pre-Roman) finds of all multiple artefacts, made of any metal, found together and single artefacts deliberately containing any quantity of precious metal. The act incorporates the former Treasure Trove, so that any objects or coin hoards less than 300 years old, made substantially of gold and silver that have been deliberately hidden with the intention of recovery and for which the owner is unknown, also constitute treasure.

The Act applies to objects found anywhere in England, Wales and Northern Ireland, including in or on land, in buildings (whether occupied or ruined), in rivers and lakes and on the foreshore (the area between mean high water and mean low water) providing the object does not come from a wreck.

In Scotland, all ownerless objects belong to the Crown. They must be reported regardless of where they were found or of what they are made. The finder receives market value as long as no laws have been broken. Not all finds will be claimed.

Here are three sample hoards associated with metal detecting.

Part of the Staffordshire Anglo-Saxon gold and silver hoard

(Photo: David Rowan, Birmingham museum and art gallery CC BY 2.0
http://commons.wikimedia.org/w/index.php?curid=7894515)

Metal Detecting Benefits for Landowners

Part of the Hoxne Roman gold and silver hoard in the British Museum

(Photo: Mike Peel, www.mikepeel.net/ CC BY-SA 4.0
http://commoms.wikimedia.org/w/index.php?curid=10754225)

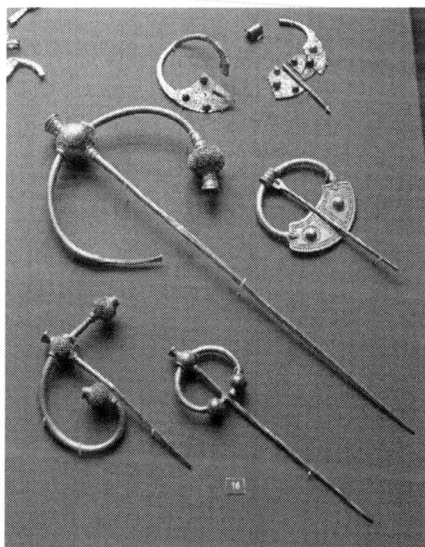

Penrith Viking silver hoard in the British Museum

(Photo: Ealdgyth CC BY-SA 3.0
http://commons.wikimedia.org/w/index.php?curid=11084188)

4 HOW WE CAN HELP YOU

Farm security

Metal detectorists act as free scarecrows. Many farmers have said that few wild birds land on fields, even neighbouring fields, where a detectorist operates, which means less seeds lost when fields are drilled. Rabbits too are deterred by metal detectors.

The presence of metal detectorists on your land can also prevent theft, trespassing, fly-tipping, vandals and other illicit or unwanted human activity. As well as, at least, providing warning of sick, injured or straying livestock.

Search and recovery service

Metal detectorists are always willing to search for lost items. This is a free service offered within reasonable geographic distance of the detectorist or a modest fee to cover travelling expenses can be negotiated. Detectorist clubs are well placed to supply experienced detectorists near your location.

We have all at some time lost something and have had to come to terms with the fact that we probably have lost that item for good.

Detectorists can help you find jewellery, keys, pieces of farm machinery and tools and also help to locate hidden services such as cables, pipes, stopcocks and drain covers.

Clearance

As well as looking for historical artefacts, the metal detectorist also provides an invaluable unwanted metal clearance task. During the search, a lot of iron, aluminium and other rubbishy items may also be dug up. Detectorists will take these away, wherever possible and dispose of them responsibly for you. In the unlikely event that anything emerges that is too large or dangerous to remove, it will be left out of harm's way or suitably marked. You or your manager will be made aware of the situation so that it can be dealt with accordingly.

Reward

You could be lucky and reap the benefits of a find of a lifetime, or gain some knowledge of the history of your land. Generally you will enjoy a free and mutual partnership with detectorists. You will be shown what has been found on your land and any particular non-treasure artefacts or coins you would like to keep will be handed over to you as is your right.

There is a legal obligation to report potential treasure finds to the coroner. Items claimed by the coroner as treasure earn a monetary award equivalent to market value, which is shared equally between the landowner and the finder unless, a prior agreement is in place. If treasure is disclaimed then it is returned to the finder, with agreement of the landowner and can be treated as any ordinary find. That is to say the find can either be shared in some way or sold and the proceeds shared.

5 DAY RATES FOR DETECTING RALLIES

As landowner you can choose whether or not to allow metal detectorists on your land. If you decide to allow detecting, you have a number of choices open to you. You might be happy to work with one or two detectorists or allow club groups on your land. The advantage of allowing groups to hold a metal detecting rally is the fee the club will pay for the event.

Rallies can be small or large, local or county and half day to full day in duration. Below is a guideline of rates you can expect to receive. All invited detectorists will carry suitable public liability insurance arranged by a national metal detecting organization.

RALLY TYPE	No. DETECTORISTS	ACREAGE*	PAYMENT
Local (half-day)	10-20	5+	£50-£100
Local (full-day)	10-20	10+	£100-£200
County	50-100**	50+	£500-£1000

* The acreage of a proposed rally site is a guideline only. For smaller sites the number of detectorists can be reduced, and/or the event repeated a number of times.

** For those of you who have large sites available for County rallies, the number of detectorists can be more than 100. If you wish to hold one of these much larger events the organizers will discuss and arrange any required facilities, such as overnight camping, with you

6 WHAT COULD BE FOUND ON YOUR LAND?

Whether you own a paddock or large garden, a farm, several farms or an estate, there could well be interesting artefacts waiting to be found.

Pasture land can hold many hidden treasures undisturbed for many years, some of which could be very old. The gardens and surrounding land of old houses or mills will undoubtedly have old coins and artefacts.

Cultivated land can yield hidden finds after each ploughing. More hoards are coming to light in recent years as metal detector technology improves and farmers have the ability to plough deeper into the soil.

The age of the land and previous human activity can be an indication of a potentially higher rate of finds. Perhaps you have an ancient stream or river running through your property or trees or buildings which are hundreds of years old. Detectorists also like to research the land they are searching. It might be worth looking in the archives for ancient activity on or around your land. You may discover that your land once included a Roman settlement or a busy market place, or a battle site or a mill...

Examples of finds

7 RESPONSIBLE METAL DETECTING CODE OF PRACTICE

A Code of Practice for Responsible Metal Detecting in England and Wales has been agreed by all key archaeological bodies, together with metal detecting and landowners' organizations. This is the first time that these bodies have joined together to precisely define responsible metal detecting and the Code provides a clear and unambiguous definition of what constitutes good practice. The code is also generally applicable to countries outside England and Wales.

The signatories are the National Council of Metal Detecting, the Federation of Independent Detectorists, the Country Land and Business Association, the National Farmers Union, the Council for British Archaeology, English Heritage, National Museums and Galleries of Wales, Museums, Libraries and Archives Council, The British Museum, the Portable Antiquities Scheme, the Society of Museum Archaeologists and the Royal Commission for the Ancient and Historical Monuments of Wales. The agreement is mainly voluntary but has the full endorsement of the signatories, and all parties are committed to ensuring its members abide by the advice set out in the document

Code of Practice for Responsible Detecting in England and Wales [annotated]

<u>Being responsible means:</u>

* *Before you go metal-detecting*

1. Not trespassing; before you start detecting obtain permission to search from the landowner/occupier, regardless of the status, or perceived status, of the land. Remember that all land has an owner. To avoid subsequent disputes it is always advisable to get permission and agreement in writing first regarding the ownership of any finds subsequently discovered (see http://www.cla.org.uk / http://www.nfuonline.com).

[A sample search agreement is included on page 28]

[An occupier of land such as a tenant farmer may need to clear any permissions with the landowner first, else risk being in breach of their tenancy agreement].

2. Adhering to the laws concerning protected sites (e.g. those defined as Scheduled Monuments or Sites of Special Scientific Interest: you can obtain details of these from the landowner/occupier, Finds Liaison Officer, Historic Environment Record or at http://www.magic.gov.uk). Take extra care when detecting near protected sites: for example, it is not always clear where the boundaries lie on the ground.

[Metal detecting is not permitted on archaeological sites on holdings with Higher Level Stewardship agreements or Sites of Special Scientific Interest (SSSI) without the permission of Natural England: such sites will be identified in the Farm Environment Plan].

3. You are strongly recommended to join a metal detecting club or association that encourages co-operation and responsive exchanges with other responsible heritage groups. Details of metal detecting organizations can be found at http://www.ncmd.co.uk / http://www.fid.newbury.net.

4. Familiarising yourself with and following current conservation advice on the handling, care and storage of archaeological objects (see http://www.finds.org.uk).

While you are metal-detecting

5. Wherever possible working on ground that has already been disturbed (such as ploughed land or that which has formerly been ploughed), and only within the depth of ploughing. If detecting takes place on undisturbed pasture, be careful to ensure that no damage is done to the archaeological value of the land, including earthworks.

6. Minimising any ground disturbance through the use of suitable tools and by reinstating any excavated material as neatly as possible. Endeavour not to damage stratified archaeological deposits.

7. Recording findspots as accurately as possible for all finds (i.e. to at least a one hundred metre square, using an Ordnance Survey map or hand-held Global Positioning Systems (GPS) device) whilst in the field. Bag finds individually and record the National Grid Reference (NGR) on the bag. Findspot information should not be passed on to other parties without the agreement of the landowner/occupier (see also clause 9).

8. Respecting the Country Code (leave gates and property as you find them and do not damage crops, frighten animals, or disturb ground nesting birds, and dispose properly of litter: see http://www.countrysideaccess.gov.uk).

After you have been metal-detecting

9. Reporting any finds to the relevant landowner/occupier; and (with the agreement of the landowner/occupier) to the Portable Antiquities Scheme, so the information can pass into the local Historic Environment Record. Both the Country Land and Business Association (http://www.cla.org.uk) and the National Farmers Union (http://www.nfuonline.com) support the reporting of finds. Details of your local Finds liaison Officer can be found at http://www.finds.org.uk, email info@finds.org.uk or phone 020

7323 8611.

10. Abiding by the provisions of the Treasure Act and Treasure Act Code of Practice (http://www.finds.org.uk), wreck law (http://www.mcga.gov.uk) and export licensing (http://www.mta.gov.uk). If you need advice your local Finds Liaison Officer will be able to help you.

11. Seeking expert help if you discover something large below the ploughsoil, or a concentration of finds or unusual material, or wreck remains, and ensuring that the landowner/occupier's permission is obtained to do so. Your local Finds Liaison Officer may be able to help or will be able to advise of an appropriate person. Reporting the find does not change your rights of discovery, but will result in far more archaeological evidence being discovered.

12. Calling the Police, and notifying the landowner/occupier, if you find any traces of human remains.

13. Calling the Police or HM Coastguard, and notifying the landowner/occupier, if you find anything that may be a live explosive: do not use a metal-detector or mobile phone nearby as this might trigger an explosion. Do not attempt to move or interfere with any such explosives.

8 SAMPLE SEARCH AGREEMENT

LANDOWNER/SEARCHER AGREEMENT

The following terms and conditions are agreed between landowner and searcher:

The landowner grants permission to the searcher to use location equipment and hand tools to search and extract finds from the ground of land known as:

....

....

During the period: **From:**............................... **To:**...

The searcher enters the land at the searcher's own risk.

The searcher shall report all worthwhile finds to the landowner within a reasonable time of being found in accordance with the landowner's wishes.

The searcher shall report any bombs, missiles or live ammunition discovered, to the landowner and to the police.

Archaeological discoveries will be reported to the landowner in the first instance. The information will then be passed on to the local museum or archaeological body providing the landowner agrees.

Potential treasure discoveries will be reported to the landowner in the first instance providing this can be achieved within fourteen days. In any event the Coroner will be informed within fourteen days as prescribed by The Treasure Act.

All finds (or the value thereof) and treasure awards will be shared equally between the searcher and landowner.

The searcher shall take great care to: work tidily, avoid hindrance to the working of the land and avoid damage to the landowner's, property, animals or crops. In the unlikely event of damage the searcher shall rectify the damage at the searcher's own expense.

The searcher shall comply with any special conditions, recorded overleaf.

This agreement may be terminated by the landowner at any time and if so terminated the searcher shall immediately cease all operations.

	SEARCHER	**LANDOWNER**
SIGNATURE:	....	....
NAME:	....	....
ADDRESS:	....	....
	....	....
	....	....
DATE:	....	....

Printed in Great Britain
by Amazon